ISBN 978-1-5277-3656-6
PIBN 10885223

1 MONTH OF
FREE
READING

at
www.ForgottenBooks.com

By purchasing this book you are eligible for one month membership to ForgottenBooks.com, giving you unlimited access to our entire collection of over 1,000,000 titles via our web site and mobile apps.

To claim your free month visit:
www.forgottenbooks.com/free885223

English
Français
Deutsche
Italiano
Español
Português

www.forgottenbooks.com

Mythology Photography **Fiction**
Fishing Christianity **Art** Cooking
Essays Buddhism Freemasonry
Medicine **Biology** Music **Ancient
Egypt** Evolution Carpentry Physics
Dance Geology **Mathematics** Fitness
Shakespeare **Folklore** Yoga Marketing
Confidence Immortality Biographies
Poetry **Psychology** Witchcraft
Electronics Chemistry History **Law**
Accounting **Philosophy** Anthropology
Alchemy Drama Quantum Mechanics
Atheism Sexual Health **Ancient History**
Entrepreneurship Languages Sport
Paleontology Needlework Islam
Metaphysics Investment Archaeology
Parenting Statistics Criminology
Motivational

UNIVERSITY OF ILLINOIS

_1344_
_/. ...../_

June. 1. .1907. 190

THIS IS TO CERTIFY THAT THE THESIS PREPARED UNDER MY SUPERVISION BY

Floyd Everett Dougherty

ENTITLED ...... DESIGN FOR A COLD STORAGE AND ICE PLANT...

IS APPROVED BY ME AS FULFILLING THIS PART OF THE REQUIREMENTS FOR THE

DEGREE OF . .Bachelor of Science in Architecture

John N. Lease
Instructor in Charge.

APPROVED: _N. Clifford Ricker._

HEAD OF DEPARTMENT OF_ Architecture

DESIGN FOR
A COLD STORAGE AND ICE PLANT

BY

FLOYD E. DOUGHERTY

THESIS

For the Degree of Bachelor of Science
in Architecture

College of Engineering
University of Illinois

PRESENTED, JUNE, 1907

1307
Ⅱ74

-A COLD STORAGE PLANT.-

      Cold storage buildings are used for the preservation
of perishable goods in a low temperature, which prevents decay.
The goods are placed in rooms, and are there kept sometimes
for months, separate rooms being usually necessary for different
kinds of goods, not only because different substances often act
on and effect one another, but also because, as experience
shows, they require different temperatures for their preserva-
tion.

      For the proper preservation of goods, it is necessary
(1) that the air should be often renewed; (2) that the air
should have the proper amount of moisture; (3) that the temperature should remain within certain limits.

      The first requisite is obtained by a proper system
of ventilation, the second by a careful use of the hygrometer
and psychrometer for ascertaining the relative humidity of the
air, care being taken not to have the air too dry, as this may
result in the evaporation of the goods, nor too damp, as this
will cause mold or mustiness. The amount of moisture that
air can absorb increases more rapidly than the temperature.
When the air enters the room it is frequently very damp, and a
part of the moisture will be precipitated on the pipes as the
temperature falls, where it will freeze. The temperature is
controled by the refrigerating machine. The brine-circulation
system has in general been found somewhat superior to the direct expansion system. The freezing rooms are placed on the

172916

ground floors, for to locate them above the cold storage rooms, even with the best insulation, the ceiling of the cold storage rooms will sweat.

The purpose of insulating a cold storage freezing room is to prevent it from receiving heat, either by radiation, by conduction or by connection. Storage rooms are insulated with wood and paper, by air-spaces filled with some non-conducting material, such as saw-dust, planing-mill shavings, mineral wool, cork, wood-ashes, cinders, etc.

The best and cheapest non-conductor is air, but in order to make this efficient, it is necessary to make a "dead air space," that is, to so enclose the air on all sides as to prevent its circulation; otherwise heat there would be conveyed by connection from the exterior into the rooms. For this reason, brick walls are pitched or covered with paraffine, and when wood is used instead of brick, paper is laid between boards so as to prevent the escape of the air confined in the air spaces.

In constructing an air-space with boards, it is best to use double boards with paper between them, this double boarding with paper making it almost impossible for the air to pass. Such construction, however, is comparatively expensive, and a single thickness of board and an air-space filled with some good non-conducting material answers the purpose. In choosing a filling material, the points to be considered, are the following: (1) That it should be a good non-conductor; (2) That it should not be too heavy, as then it is liable to settle

and leave a portion of the top insulation unprotected; (3) that it should not be affected by dampness to any appreciable extent, and (4) that it should not be too expensive.

The best non-conductor and insulator for filling an air space is granulated cork, but this, being very expensive, is seldom used; it is light, is not affected by dampness, and consequently does not mold or rot. Mineral wool forms an excellent non-conductor, and if the best quality is purchased it is quite light, only weighing about seven pounds per cubic foot, and it does not readily settle down but is easily affected by moisture, and it then becomes quite soggy and settles rapidly.

Where sufficient headroom is available, it is best to put horizontal strips half way up the filling in air spaces, so as to divide the filling into two sections. This divides in half the weight of the filling and prevents its settling. For practical purposes, planing mill shavings in bales can be bought at very reasonable rates.

They are thoroughly dry and are little affected by dampness. Planing-mill shavings are much better than saw-dust, being a poorer conductor, less easily affected by moisture, and less liable to settle.

Spruce is found to be the most satisfactory wood for insulating cold storage rooms, being free from the odor of white or yellow pine.

Exposed pipes or other surfaces through which a refrigerating liquid is flowing will soon be conveyed with frost. This is because the moisture in the atmosphere is

deposited on the pipes and is then frozen.

To thoroughly insulate such surfaces, it is not only necessary to keep the heat from passing through them, but to keep the air out or make the insulation perfectly air tight. Hair felt is wrapped around the pipe, and then covered with paper, then more felt and more paper, and so on for several layers, when the whole is sewed in canvas and covered with several coats of good water proof paint.

The first requirement of a cold storage room being good insulation, all windows and other openings by which light is admitted to the room should be closed with light-tight shutters, so as to prevent any daylight from entering. If this is not done, radiant heat will enter the room. Such windows can be closed with large shutters on the inside, similar to ice-house doors, which can be opened for ventilation when desired. A cold storage room should be provided with some means of ventilation, so that the rooms can be thoroughly aired and the air renewed.

There are three methods employed for maintaining the temperatures of cold storage and freezing rooms. The first is by means of direct radiation, where the brine pipes are run in the room and the brine is allowed to circulate through them. The second is by indirect radiation, where the pipe coils are placed in a coil bunker on the upper floor and the air is allowed to descend to the floors below by gravity, and after absorbing heat it is returned to the coil bunker by means of flues or ducts for that purpose.

The third and most approved method is by means of a
fan or blower, which sets the air in circulation and controls
its distribution.

-ICE PLANT-

The apparatus used in the can system consists of a
large rectangular wood or iron tank containing the expansion
coils or pipes. Galvanized iron cans are placed between expansion coils. These cans are filled with distilled water,
and when the brine is chilled below the freezing point, the
water in the cans freeze. If the temperature of the brine is
not allowed to fall below 25^0 and ordinary well water is used
in the cans, the ice produced will be comparatively clear on
the outside and somewhat granulated in the center. If, however,
the brine temperature is allowed to fall about to 15^0 the ice
will be entirely opaque.

To get good, clear ice, distilled water is used. This
makes a bright, clear ice, with the exception of a small core
or feather in the center. It is necessary to have a good
distilling apparatus.

The distilled water is usually made by condensing the
exhaust steam of the engine operating the compressor in case of
a compression plant, or by cooling the condensed steam that
leaves the generator.

Ordinarily the steam required by a first-class machine
of either kind is considerably less in weight than the amount
of ice that the machine can make in a given time. It is therefore necessary to draw live steam from the boiler and condense

it, in order to make up the deficiency. It follows from this that the economy of the plant depends entirely upon the economy of the boiler.

Ice tanks are usually made of steel, though in some plants wood is used. Steel tanks should be made of $\frac{1}{4}$ inch steel for tanks 3 feet or more in depth. The tank should be properly braced and reinforced and have an angle iron rim punched for bolt-holes around the top, so that the grating can be securely attached thereto.

Expansion coils should be of extra heavy pipe and should run the full length of the tank, one coil between each row of cans.

The coils should be continuously welded throughout their lengths with tails carried through the end of the tank, stuffing-boxes being provided on the tank to prevent brine leakage. The coils are usually strapped in such a manner as to permit the grating to be supported by the straps. There should be 100 square feet of coil surface for each ton of ice made. Each coil should be provided at its inlet with an expansion valve and the outlet should be provided with a stop-valve.

The brine tank, should be set on a brick foundation. On top of this foundation, the insulation for the bottom of the brine tank is built in the form of a floor. The best thing to use for the filling between the joists is granulated cork, but this is very expensive, and planing mill shavings are often used instead. Care should be taken not to use sawdust, owing to the possibility of spontaneous combustion in case it becomes

dampened, and because of its avidity for moisture. A well insulated brine tank consists of the main joists resting on the foundation, two air spaces, and one half to one inch pitch, in which the brine tank is bedded.

The insulated cover of the brine tank need not be as heavily insulated as the bottom and sides.

Ice plants are usually provided with hand cranes of the traveling type. A crane of this kind consists of a light channel iron frame on which moves a four wheel trolley provided with a geared hoist. On the drum of the hoist is run a rope or chain and to this is fastened a can latch. The latches are provided with hooks on the under side; these hooks engage into the holes in the sides of the can, so when the ends of the latches are moved the hooks disengage and the can releases.

After the ice is dumped out of the can, the latter is brought back to its original place in the tank. It is necessary to refill it with water for freezing. This accomplished by means of the distilled water system, and to this hose the can filler is attached.

After the can is placed in its position in the ice tank, the can filler is inserted and a trigger with which it is provided is pulled. This starts the water running into the can. When the water has reached the desired level in the can, a ball float strikes the trigger and shuts off the water supply. In this manner, any number of ice cans may be filled, the operation requiring little attention. All the cans are filled to exactly the same level.

The walks of the ice storage should be furred out
to a distance of 6 or 8 inches with slats on the furring pieces
to prevent the ice from coming into contact with the warm
walks. These rooms should be kept at a temperature of about
28°, no lower temperature being advisable, as the ice is then
liable to check or honey-comb. The ice should be set on end
with a space of about one half inch left between the cakes,
and slate or boards should be placed over the first layer before
the second is placed.

The temperature of the ice storage room should not
be allowed to exceed 30°, as the ice is then liable to melt,
and if this happens, the cakes will freeze together when the
room cools down again and it will be necessary to quarry out
the ice.

General Calculations for Ice Plant.

The plant is to have a capacity of 100 tons of ice
during each 48 hours, containing 1000 cans, the ice in each can
weighing 200 lbs., condensing water taken at 70°, Fah.

Ice machines using ammonia at about 190 lbs., per sq.,
in.,above atmospheric condensing pressure and 15 lbs., suction
pressure.

Under the summer conditions with condensing water
at 70° F., the compression machine operates with 190 lbs.,
guage condenser pressure and 15 lbs., guage suction pressure.
In this type of machine the useful circulation, allowing for
cylinder heating, is about 13 lbs., of ammonia per hour indi-

cated steam engine, horse power. This weight of ammonia will produce 32 lbs., of ice at 15^{0} F., from water at 70^{0} F.,

In order to compensate for losses of steam from the boiler leaks, etc., the quantity of steam required is taken at 33% in excess of that theoratically required to produce the ice. The total steam per horse-power will therefore be 32 X 1.33 = 43 lbs. If we assume an evaporation of $8\frac{1}{3}$ lbs., of water per pound of coal, the amount of ice produced per one pound of coal would be 6 lbs. If for every ton of ice made, 400 lbs., of water are wasted by leakages, 2400 lbs will have to be distilled for making one ton of ice. If $3\frac{1}{2}$ indicated horse-power is required per ton of ice made per 24 hours though it takes 48 hours to freeze, then 1 3/4 horse-power will be required per ton of ice. For 100 tons of ice per 48 hours the plant will require a 175 horse-power compressor.

Using the "York and St. Clair" compound ice machines rated at 50 to 55 tons per 24 hours, a 225 H. P. boiler is needed with a grate area of 75 sq., ft., requiring a chimney 43 in., in diameter and 75 feet high.

There is 175,00 cubic ft., of ice storage to be kept at 10^{0} F.. One running foot of two inch pipe will suffice for 10 cu., ft., of space with brine circulation, this requiring 1750 running feet of two inch pipe, 250 feet of 2" pipe is required for freezing per ton of ice every 24 hours, or 175 feet per ton every 48 hours, thus requiring 17500 feet of two inch pipe.

The temperature of cold rooms are to be kept at

32° to 35°, this will require 1 running foot of 2" pipe per each 20 cu., ft.,of space in cold rooms. There being a total of 333,090 cu. ft., of storage rooms, 16,655 feet of 2 inch pipe will be required, this giving a total of 35,905 running feet of two inch pipe.

SUMMARY.

Chimney 43 inches in diameter, 75 feet high.
Grate area 75 sq., ft.,
Boilers H. P. 225

Compressire H. P. 175.

Ice Making---running feet of 2" pipe					17500	feet
10° storage	"	"	"	"	"	1750	"
32° "	"	"	"	"	"	16655	"
				Total		35905	

CPSIA information can be obtained
at www.ICGtesting.com
Printed in the USA
BVHW081059110219
539956BV00021B/2634/P

9 781527 736566